WITHDRAWN

BERMUDA
TRIANGLE

BERMUDA TRIANGLE

ODYSSEYS

KEN KARST

CREATIVE EDUCATION · CREATIVE PAPERBACKS

Published by Creative Education and Creative Paperbacks
P.O. Box 227, Mankato, Minnesota 56002
Creative Education and Creative Paperbacks are imprints of
The Creative Company
www.thecreativecompany.us

Design by Blue Design; production by Colin O'Dea
Art direction by Rita Marshall
Printed in the United States of America

Photographs by 123RF (Anton Balazh), Alamy (Cultura
Creative, FL Historical J, Grant Henderson, Sueddeutsche
Zeitung Photo), Creative Commons Wikimedia (Leonidas
Kourentis, L. Prang & Co., Boston/Library of Congress, Andrew
Moore/Flickr, NASA/Kennedy Space Center, NASA Goddard
Space Flight Center/Jeff Schmaltz/MODIS Land Rapid
Response Team, Naval History and Heritage Command/U.S.
Navy), Getty Images (Bettmann, Toronto Star Archives/
Toronto Star), iStockphoto (Adventure_Photo, gremlin,
nusuke, jpgfactory, kadrajserap, Kuzma, Valerie Loiseleux,
PetrBonek, tswinner, ursatii), Shutterstock (Vadim Sadovski)

Library of Congress Cataloging-in-Publication Data
Names: Karst, Ken, author.
Title: Bermuda Triangle / Ken Karst.
Series: Odysseys in mysteries
Includes index.
Summary: An in-depth study of the Bermuda Triangle,
examining legends, popular reports, and scientific evidence
that supports or refutes the mysterious phenomena reputed
to occur there.
Identifiers: LCCN 2019049717 / ISBN 978-1-64026-361-1
  (hardcover) / ISBN 978-1-62832-893-6 (pbk) / ISBN 978-1-
64000-515-0 (eBook)
Subjects: LCSH: Bermuda Triangle—Juvenile literature.
Classification: LCC G558.K77 2020 / DDC 001.94—dc23

First Edition HC 9 8 7 6 5 4 3 2 1
First Edition PBK 9 8 7 6 5 4 3 2 1

CONTENTS

Introduction

On August 3, 1492, Christopher Columbus left Spain in a convoy of three small ships, heading west across the Atlantic Ocean, certain he would find a new trade route to Asia. After two months, his ships were stranded in strange waters that were covered with seaweed. His uneasy sailors thought this was a sign they were near land. But land did not appear. Instead, their compasses started acting strangely, no longer pointing to where the

OPPOSITE: More than two months after leaving Spain in 1492, Columbus and his crew landed on an island in the Bahamas.

9

crew knew north to be. Then they spotted strange lights at night—a fireball that seemed to drop into the sea and a flickering in the distance that rose like a candle flame before disappearing.

When they finally found land, the crew thought they'd reached Asia, unaware they'd come to islands in what is now called the Bahamas. Nor did they know they had sailed through the Bermuda Triangle, a region of the sea that has become infamous for swallowing ships and planes and their crews without leaving a trace. Columbus made four trips through the region, turning a new page in the history of civilization. But hundreds of other travelers have simply vanished. More than 500 years later, what happened to some of them remains a mystery.

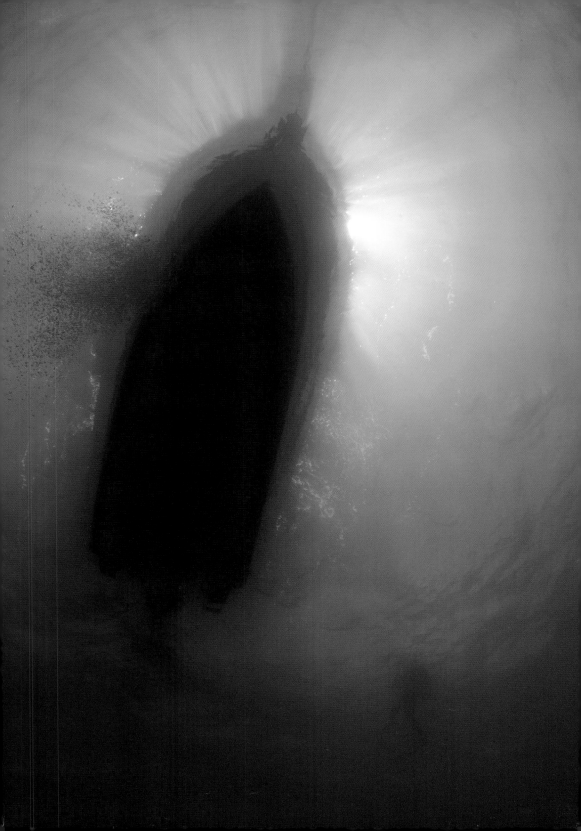

A Many-sided Triangle

It's simple geometry: A triangle is a shape with three corners and straight sides. But nothing is simple about the Bermuda Triangle. For one thing, the Bermuda Triangle is not labeled on any maps. That's because the United States government and others that make maps don't acknowledge its existence. But it might also be because the Bermuda Triangle can assume many shapes. It's a different

OPPOSITE: The deepest point in the Atlantic Ocean is the Milwaukee Deep (or Depth), located in the Puerto Rico Trench, which lies within the Bermuda Triangle.

place to different people.

In most discussions, the Bermuda Triangle is what author Vincent Gaddis described in his 1964 article in *Argosy* magazine, in which he coined the term. It's basically the area of the Atlantic Ocean with Miami, Florida, on one corner, Bermuda on another, and Puerto Rico on the third. Hundreds of ships and planes have disappeared in that region in the last two centuries. The area outlined by those corners covers 500,000 square miles (1,294,994 sq km), an area slightly bigger than California, Oregon, Washington, and Montana put together.

Since Gaddis's article, though, the legend of the Bermuda Triangle has sometimes loomed even larger. Some have described the borders as forming an irregular shape that would include much of the southeastern coast of the U.S., from the Florida Keys to as far north

TAKEAWAY

This much larger area has been called the "Limbo of the Lost," a term first used by author John Wallace Spencer in 1973.

as Wilmington, North Carolina; then it would extend east beyond Bermuda, south into the West Indies, and northeast to include the waters around Cuba, Haiti, and Jamaica, coming close to Mexico's Yucatan Peninsula. Some incidents attributed to the Bermuda Triangle occurred outside even those generous borders, approaching the Azores, a cluster of islands in the Atlantic that is far closer to Europe and Africa than to Florida, Puerto Rico, or Bermuda. This much larger area has been called the "Limbo of the Lost," a term first used by author John Wallace Spencer in 1973 and later the title of an adventure video game.

The Bermuda Triangle, and the sea around it, has come to be known by other names as well: the Hoodoo Sea, the Sea of Oblivion, the Sea of Lost Ships, the Graveyard of the Atlantic, and the Sea of Fear. The last three names in particular have been attached to the Sargasso Sea, a large part of the Atlantic that includes a broad expanse of the Bermuda Triangle. It was there that Columbus was idled amid endless blankets of seaweed (the word "Sargasso" comes from the Portuguese word *sargaço*, meaning seaweed). Columbus evidently wasn't the first to experience such entrapment. As long ago as 500 B.C., an admiral named Himlico, from the ancient North African city of Carthage (in what is now Tunisia), wrote of seaweed in the Atlantic that "holds back the ships like bushe" and where "monsters of the sea move continuously to and fro, and fierce monsters swim among the sluggish and slowly creeping ships."

Modern science has helped explain why the Sargasso might have seemed like a pool of evil forces to ancient mariners. Three major ocean currents—the Gulf Stream and the North Atlantic Drift flowing from the Gulf of Mexico, northwards along the coast, across to northern Europe, and the North Equatorial Current, flowing west from Africa to North America—run clockwise around the edges of the Sargasso, leaving a huge, calm region between them that collects seaweed and debris from several continents. Long ago, though, sailors simply regarded it as a ghostly place, where lost ships from sailing cultures (purportedly from the Vikings to the Spanish, and possibly even those from the mythical lost continent of Atlantis) would drift endlessly, with crews that had become skeletons. Some tales had the ships being slowly dragged beneath the surface by seaweed that grew over them like a net. Others suggested it was a place where horses, tossed overboard by ships' crews in desperate attempts to save drinking water while caught in the windless latitudes, frolicked in spirit form.

If Columbus believed any of those accounts, he certainly faced down his fears. He sailed four round trips through the Sargasso and on through the Bermuda Triangle. Ships that followed Columbus in the next several centuries disappeared with some regularity, thanks to hurricanes, piracy, mutiny, smuggling, and other dangers. Many wrecks, of course, were found on the bottom of the sea. Others were never found, but because communication, record keeping, and transportation weren't as sophisticated as they are now, it wasn't exactly a mystery.

Those things just happened sometimes. It could have been anything.

But in 1800, a U.S. Navy schooner named the *Pickering* disappeared on a voyage from Boston to the West Indies. Did she go down in a gale? No one knows. No wreckage and no crew members were ever found, which seemed suspicious considering that ship traffic was heavy along the East Coast. Someone should have seen something. For that reason, the *Pickering* is regarded as the first of the modern Bermuda Triangle mysteries. Today, those mysteries feature several common elements: reports of good weather and easy travel followed by reports of mysterious fogs and suddenly appearing clouds, spinning compasses, and gaps in time, and then a complete disappearance. Since the *Pickering* incident, hundreds more ships and, later, planes vanished off the southeastern U.S.

Translating the Triangles

It's too bad Charles Berlitz never encountered any aliens, because he might have spoken their language. Berlitz, the grandson of the founder of Berlitz Language Schools, spoke eight languages by the time he was a teenager. He studied at Yale University and then served with the U.S. Army in counterintelligence, which tries to protect government secrets and other information by preventing others from stealing it. His interest in diving and archaeology led him to write a popular book, *The Mystery of Atlantis*, in 1969, and to follow that up with *The Bermuda Triangle: An Incredible Saga of Unexplained Disappearances*, which sold 20 million copies. Berlitz argued that the Bermuda Triangle is a one-way passage to another dimension through which humans cannot return. He linked the sagas of Atlantis and the Bermuda Triangle, writing, "The Bermuda Triangle leads back to lost and sunken lands, to forgotten civilizations, to visitors to the earth through the centuries from outer or inner space whose provenance [origin] and purpose are unknown." Another Berlitz book, *The Roswell Incident*, was one of the first to assert that the U.S. government was hiding aliens that had landed in a spacecraft in New Mexico in 1947.

TAKEAWAY

Gaddis wrote that, from the 1940s to the '60s, nearly 1,000 people had vanished in the Bermuda Triangle.

coast. That gave rise to the idea of the Bermuda Triangle as a place where travelers were not safe. Gaddis wrote that, from the 1940s to the '60s, nearly 1,000 people had vanished in the Bermuda Triangle—*totally* vanished, without so much as a hat left floating or washing up on shore. Entire ships, sailboats, and planes, some while within view of shore or reportedly on their way in for a landing—were said to have disappeared, *poof!*—without even an oil slick or life preserver left behind. The most prominent incident was Flight 19, a group of five small navy bombers that never returned from a practice run over the Atlantic from Fort Lauderdale, Florida, in

1945. A search plane sent out after the squadron also disappeared shortly after taking off. In all, 6 planes and 27 men were lost, but not a trace of them or their aircraft was ever found.

Author Gian J. Quasar, in his 2005 book *Into the Bermuda Triangle*, says official records show the losses to be about 4 planes and 20 yachts per year. Is that cause for alarm? A 1974 report for the U.S. Naval History and Heritage Command, by Howard L. Rosenberg (who went on to become a producer for the television news program 60 *Minutes*), noted that the U.S. Coast Guard answered 8,000 distress calls in the area in 1973 alone. Meanwhile, about 150,000 ships, boats, and planes were crossing safely through the area each year, far outweighing the number of documented disappearances. Lloyd's of London, an insurance company, reportedly did not regard travel through the area as being risky for its clients. Armed with such facts, and possible explanations for the disappearances, Rosenberg wrote that Bermuda Triangle mysteries were "a voluminous mass of sheer hokum."

However, Gaddis's words amplify the legend of the Bermuda Triangle. "Its history of mystery dates back to the never-explained, enigmatic [puzzling] light observed by Columbus when he first approached his landfall in the Bahamas," Gaddis wrote. "The Bermuda Triangle underlines the fact that despite swift wings and the voice of radio, we still have a world large enough so that men and their machines and ships can disappear without a trace."

Time Warps?

There is no shortage of explanations for why ships and boats have disappeared in the Bermuda Triangle. Some theories are based on reasonable speculation involving normal **phenomena**—sudden bad weather, human error, or mechanical problems—with crash remains being quickly carried away by fast-moving ocean currents. Many researchers say that people who have written about the haunting stories in the area have

OPPOSITE: Some people think advanced technology from the legendary lost civilization of Atlantis causes problems for passing ships and planes.

simply overlooked or ignored evidence of such common problems. But of course there are other possibilities. Some say aircraft that vanish from the Bermuda Triangle have simply flown through a kind of time warp into another dimension. Ships might have been caught in an upwelling of methane from the depths, which would make the surface less dense, causing anything on top of the water to sink suddenly. Some suggest that an energy disruption coming from beneath the sea, perhaps originating from the lost continent of Atlantis, could be to blame. Another idea is that vessels have been captured by aliens that are interested in humans. Similarly, some writers have suggested there could be an alien civilization beneath the sea that, after existing quietly for ages, has become active out of concern that pollution may be destroying its environment. Yet another notion is that those deep-sea

The U.S. Coast Guard's position is that it is "unimpressed with supernatural explanations of disasters at sea."

beings are trying to investigate nuclear power—accessible on some submarines—to somehow prevent humans from misusing it. And then there are the official statements.

The U.S. Coast Guard's position is that it is "unimpressed with supernatural explanations of disasters at sea. It has been our experience that the combined forces of nature and unpredictability of mankind outdo even the most far-fetched science fiction many times each year." Furthermore, the National Oceanic and Atmospheric Administration (NOAA) said there is "no evidence that mysterious disappearances occur with any greater frequency in the Bermuda Triangle than in any other large, well-traveled area of the ocean." But a coast guard officer, speaking to naval investigators after the loss of Flight 19 and the search plane in 1945, had a different outlook: "We don't know what the heck is going on out there," he told them candidly. An officer of the Naval Board of Inquiry also described the incident as baffling. The navy bombers, he said, "vanished as completely as if they had flown to Mars."

Some say that phrase helped push the Bermuda Triangle into the realm of the extraterrestrial. In the first few years after that squadron of bombers and the search plane were lost, the incident was treated as an isolated event, although it remained puzzling. But in his 1964 article, Vincent Gaddis made a case for "a mysterious menace that haunts the Atlantic off our southeastern coast" and

cautioned that "it may strike again." He detailed nearly 20 incidents of ships and planes that had vanished in the area, going back to 1840. But he offered essentially one "possible theory" that would have affected only planes: the aircraft had encountered "a hole in the sky," which he described as "an unknown type of atmospheric aberration [unusual occurrence]."

Ivan T. Sanderson, science editor at *Argosy*, the magazine in which Gaddis's article appeared, was also well known as a Bigfoot researcher. He developed a theory that the Bermuda Triangle was one of 12 triangular regions, evenly spaced around the globe, where colliding ocean currents set up vortices that disrupted radio signals, magnetic fields, and perhaps even gravity itself. His idea tied in with that of Wilbert Smith, a Canadian researcher who, while studying unidentified flying objects (UFOs)

Sanderson developed a theory that the Bermuda Triangle was one of 12 triangular regions, evenly spaced around the globe.

in the early 1950s, developed a theory that there were small areas on Earth where magnetic forces were violently unstable. Disturbances occurred without warning and often altered the Earth's magnetic field. This became a popular idea among Bermuda Triangle researchers and writers, who seized on it as something that would have accounted for the spinning or malfunctioning compasses often reported in Bermuda Triangle disappearances.

Several pilots have introduced the idea of an "electronic fog," based on their experiences of being enveloped in foggy conditions that didn't appear on radar. The fog seemed to cause their instruments to go haywire,

In Fact, It Was Fiction

In 1884, a London magazine published the short story "J. Habakuk Jephson's Statement," which claimed to be an eyewitness account of a bloody incident at sea 12 years before. In the telling, Jephson had been the doctor aboard a ship called the *Marie Celeste*, which had been boarded by African pirates off the coast of Portugal. The intruders killed the captain, crew, and passengers, sparing only Jephson, and sailed the boat to Africa. The "statement" was actually written by Arthur Conan Doyle, a young author who would soon become famous for his Sherlock Holmes mysteries. Doyle had inflated the story of the *Mary Celeste*, a boat found abandoned but intact, with its cargo of alcohol largely undisturbed. Some speculate that the captain and crew, fearing the alcohol was about to explode, got into a lifeboat tied to the ship but somehow became separated from the vessel, which continued sailing while they disappeared. Elements of Doyle's story—particularly the description of the full breakfast cooling on the table of the abandoned boat—helped the *Mary Celeste* come to be regarded over time as a Bermuda Triangle mystery, even though the boat was nowhere near the region, and the story was ultimately known to be fiction.

Several pilots have introduced the idea of an "electronic fog," based on their experiences of being enveloped in foggy conditions that didn't appear on radar.

so the pilots couldn't tell what their location, altitude, or speed were. Pilot Bruce Gernon had one of the most vivid of such foggy encounters. Flying from the Bahamas to Florida in 1970, Gernon was surprised by a quick-rising cloud that none of the air traffic controllers could see on radar. The cloud formed rings that surrounded his plane, and his compass stopped working. He kept flying his plane faster, but when he saw a clearing ahead and pushed his plane still faster, the clouds seemed to tighten around him. Finally, he escaped the cloud, and his instruments came back on line. Upon landing in

Miami, he discovered that his flight, which should have taken 75 minutes, had taken only 47. "I believe I was actually seeing the fabric of time itself," Gernon said in an interview in a 2006 episode of the *Is It Real?* series on the National Geographic Channel.

Other such dramatic stories had an effect on researcher Larry Kusche. A pilot and librarian at Arizona State University, Kusche researched dozens of Bermuda Triangle incidents going back to Columbus's time, and even flew through the fabled region himself.

He found that many of the best-known incidents had reasonable explanations and that the region's strong sea currents, frequent hurricanes, heavy winds, and high, fog-forming humidity have combined to generate strange stories for centuries. For example, reports of ships stuck in the Sargasso Sea, surrounded by seaweed with little wind, became stories of huge monsters devouring the craft. More recently, Kusche argues, the Bermuda Triangle legend has grown out of undisciplined research; incomplete, lazy, or misleading news reporting; and plain old hype. He outlined his findings on more than 50 disappearances in his 1975 book, boldly titled *The Bermuda Triangle Mystery Solved*.

The *Mary Celeste*, a brigantine found abandoned but fully outfitted in 1872, is one of the most vivid Bermuda Triangle mysteries involving ships. But Kusche notes that it was found near the Azores, more than 2,000 miles (3,219 km) east of the customary edge of the Bermuda Triangle. Its story was popularized in an anonymous 1884 article written by Sir Arthur Conan Doyle, author of the Sherlock Holmes mysteries.

The *Cyclops*, a type of ore- or coal-carrying ship known as a collier, and one of the largest ships of its day, disappeared in calm weather in 1918 on its way from Barbados to Baltimore, Maryland, according to Bermuda Triangle lore. But Kusche writes that all the conventional stories about it ignore the probability that the ship got caught in a ferocious storm with a bad engine in an area where searchers never thought it might be. And what about Flight 19, the group of five Avenger bombers that vanished on a training run in 1945, making it the most haunting of the Bermuda Triangle aviation legends? Inexperience, faulty equipment, communication breakdowns, adherence to a military discipline that overpowered some of the pilots' sense that they were flying the wrong way, and even the commander's lack of a wristwatch all led to the tragedy, according to Kusche.

Interestingly, Kusche's book also agrees that many of the incidents he investigated remain mysteries. And that only reinforces the power of the Bermuda Triangle, writes rival researcher Gian J. Quasar. Quasar asserts that Kusche exhibited the same sloppy research as those he criticizes, thereby failing to "solve" anything. So the legend lives on.

Pieces of a Legend

On December 5, 1945, five U.S. Navy Avenger bombers, in a group called Flight 19, sliced into the sky from the Fort Lauderdale, Florida, Naval Air Station, with 14 men heading east on a practice bombing run. They were led by lieutenant Charles C. Taylor, a pilot with 2,500 hours of flying experience. After dropping their bombs on a wrecked ship near Bimini, the group then headed east to practice **navigation**. Their course was to take

OPPOSITE: The U.S. Navy initially held pilot Charles C. Taylor responsible for the loss of Flight 19, but it later blamed the disappearance on "causes unknown."

Taylor seemed to think they were over the Florida Keys, and that land was to the north. Soon, all radio contact faded.

them briefly north, then back southwest directly to their base. But at about the time they should have been requesting landing instructions, Taylor reported he didn't know where they were, that he couldn't see land, and that his compasses were malfunctioning. Pilots in the other planes could be heard arguing that they should simply turn west, back toward land. But Taylor seemed to think they were over the Florida Keys, and that land was to the north. Soon, all radio contact faded. Then it was lost.

Other planes took off to search for Flight 19. One, a huge Martin PBM Mariner, was capable of landing in rough seas. With a crew of 13, it was well-suited to a major rescue operation. But after sending a radio message that all was well 23 minutes after taking off from the Banana River Naval Air Station, it, too, disappeared.

Within hours, 6 planes and 27 men had been lost in the heart of the Bermuda Triangle. Not a shred of wreckage was ever found. The disappearance of Flight 19 has become the modern emblem of the Bermuda Triangle mysteries. How could a military operation involving trained pilots, teamwork, and advanced communications devices—including radios and emergency locators—end in such an enigma, with little evidence and no answers? Throughout the centuries, there have been many other disappearances in the area involving large ships and planes. The astonishing, suspicious, and sometimes tragic stories could fill volumes, and have. The following summaries describe a few of the most dramatic tales:

The British navy training ship HMS *Atalanta* left Bermuda on January 31, 1880, bound for England. On board were 290 cadets. No trace of the vessel was ever found. Although most of its 3,500-mile (5,633 km) course would have been through seas well outside the Bermuda Triangle, the first 500 miles (805 km) were thought to be within the region.

Joshua Slocum was the first man to sail alone around the world, which he did from 1895 to 1898. Having spent most of his life at sea, he had a reputation as a sailor who could handle anything. In 1909, he set sail from the Massachusetts island Martha's Vineyard for South America and was never heard from again. Did his boat capsize? Was he hit by a waterspout? Was he overcome by some other force? No one knows.

The USS *Cyclops* was a 542-foot-long (165 m) collier that carried troops, fuel, and other materials all over the Atlantic during World War I. In February of 1918, the seven-year-old American ship (nearly new, as ships go) left Brazil with a load of manganese and a crew of 306. It stopped at Barbados to resupply on March 4 and was due in Baltimore nine days later. But after leaving Barbados, it was twice spotted heading south—the wrong direction—by British patrol boats, and it never arrived in Baltimore. The vessel's vanishing act represented

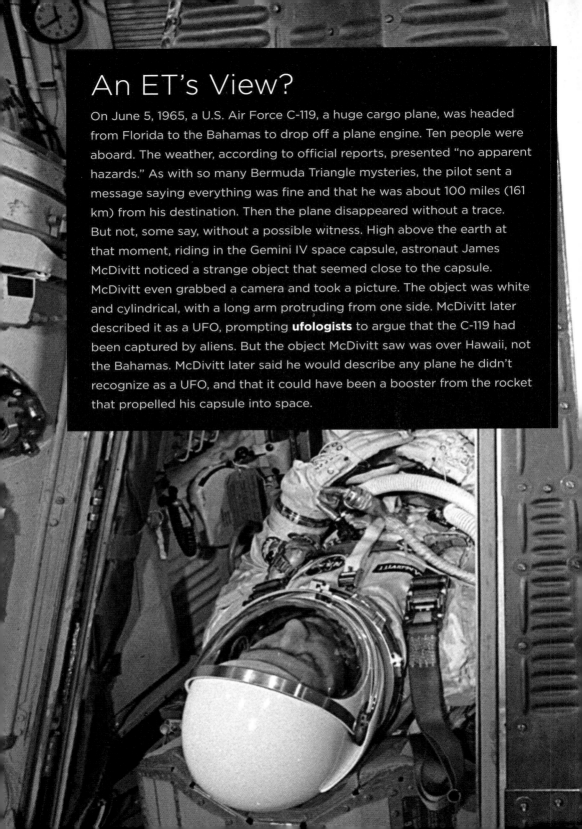

An ET's View?

On June 5, 1965, a U.S. Air Force C-119, a huge cargo plane, was headed from Florida to the Bahamas to drop off a plane engine. Ten people were aboard. The weather, according to official reports, presented "no apparent hazards." As with so many Bermuda Triangle mysteries, the pilot sent a message saying everything was fine and that he was about 100 miles (161 km) from his destination. Then the plane disappeared without a trace. But not, some say, without a possible witness. High above the earth at that moment, riding in the Gemini IV space capsule, astronaut James McDivitt noticed a strange object that seemed close to the capsule. McDivitt even grabbed a camera and took a picture. The object was white and cylindrical, with a long arm protruding from one side. McDivitt later described it as a UFO, prompting **ufologists** to argue that the C-119 had been captured by aliens. But the object McDivitt saw was over Hawaii, not the Bahamas. McDivitt later said he would describe any plane he didn't recognize as a UFO, and that it could have been a booster from the rocket that propelled his capsule into space.

the greatest loss of life in any U.S. Navy noncombat action. The boat was also the largest navy ship ever lost without a trace. The navy describes the loss as "one of the sea's unsolved mysteries," and it may be one of the most complicated. Though the ship's captain, George W. Worley, had commanded the *Cyclops* since its maiden voyage, investigations raised suspicions that, because he'd been born in Germany, he might have steered the ship deliberately into German hands. But no record of such action was ever found in German naval records. Other accounts indicate that Worley was disliked by his crew, and at the least was quite odd, often appearing at night on deck wearing long underwear and a derby hat. Still others raise suspicion about the boat itself. Could the ore it carried, known to grind against surfaces, have weakened the *Cyclops*'s hull or shifted, causing the hull to split and

Passengers aboard the luxury airliner *Star Tiger* were likely dozing as the plane neared the end of its overnight voyage from London to Bermuda on January 30, 1948.

possibly take on water? Two sister ships of the *Cyclops*, the *Proteus* and the *Nereus*, also vanished within a month of each other in November and December of 1941. The fact that they too vanished in the Bermuda Triangle linked them to the mystery surrounding the *Cyclops*.

Passengers aboard the luxury airliner *Star Tiger* were likely dozing as the plane neared the end of its overnight voyage from London to Bermuda on January 30, 1948. One account has it that the pilot, captain Brian W. Mc-Millan, flying into strong headwinds, radioed to Bermuda that the flight would be about 90 minutes late. Another

has it that he radioed saying he was only 440 miles (708 km) out, and things were going smoothly. The plane could float and carried individual life rafts, each equipped with hand-crank radios that could send emergency messages to planes and ships within 10 miles (16.1 km). But the plane vanished, and search-and-rescue teams found no trace of the aircraft or its passengers. Even Larry Kusche writes that the *Star Tiger*'s disappearance "thwarts all explanation ... It is truly a modern mystery of the air."

The story of the *Carroll A. Deering*, a five-masted schooner, is unusual in Bermuda Triangle lore. No survivors were found, but nearly everything else was, including the ship itself, with two cats on board and a meal being prepared on the stove. The boat was spotted aground by a U.S. Coast Guard watchman near Cape Hatteras, North Carolina, just after dawn on January

These incidents make up only a small fraction of the documented accounts of disappearances, abandonments, close calls, and other dramas at sea and in the air of the Bermuda Triangle.

31, 1921. The area is not in the Bermuda Triangle as it's commonly drawn but is well within the bounds according to many other versions of it. All 11 crew members and their personal belongings were gone. So were the lifeboats. Mutiny? Piracy? Bad weather? No determination was ever made. "We might as well have searched a painted ship on a painted ocean for sight of the vanished," said a federal investigator, echoing the words of Samuel Taylor Coleridge's poem, "The Rime of the Ancient Mariner."

On December 28, 1948, a Douglas DC-3 carrying 29 passengers was approaching Miami, on its way from

San Juan, Puerto Rico. Although it had some battery and radio problems upon takeoff, the pilot was able to notify air traffic controllers at 4:13 A.M. that he had a nice tailwind and that things were going well. The plane was only 50 miles (80.5 km) out, and he could see the lights of the city. But Miami didn't receive that message. Instead, a controller in New Orleans did. The message was relayed to Miami, but attempts to contact the plane were unsuccessful. The aircraft never arrived in Miami, and no trace was found.

These incidents make up only a small fraction of the documented accounts of disappearances, abandonments, close calls, and other dramas at sea and in the air of the Bermuda Triangle. As researcher Gian J. Quasar writes, "There is no other place on Earth that challenges mankind with so many extraordinary and incredible events."

New Twists

The National Museum of Bermuda doesn't have any exhibits or insights into the Bermuda Triangle, which suggests the institution regards it as a myth. But as long as planes crash and ships sink, the Bermuda Triangle is likely to continue to take some of the blame. At the very least, it has served as an explanation for the unexplainable. And that has helped it generate some spinoffs.

Several Bermuda Triangle

OPPOSITE: Wreckage of planes similar to those from Flight 19 has been found, but nothing has been able to be linked to the ill-fated Avengers.

researchers note that because of the way Earth's magnetic field works, there is a meridian on the opposite side of the world where, as with the Bermuda Triangle, compasses don't point north. This causes ships and planes to travel off course and sometimes disappear. This area, off the southern coast of Japan, is often called the Devil's Sea or the Dragon's Triangle. But Larry Kusche found that the area has long been a busy fishing ground, and many of the boats lost had been unsound to begin with, having substandard radio equipment. He also noted that the notion of the "Devil's Sea" is virtually unknown in that part of the world.

Nevertheless, there now may be a new Bermuda Triangle, not far from the old one. In recent years, a number of plane crashes and disappearances just south of the Bermuda Triangle, off the coast of Venezuela,

There have been more than 15 reported incidents of small planes crashing, vanishing, or declaring emergencies while flying in the area.

have given rise to "Los Roques Curse." The name comes from a cluster of islands about 440 miles (708 km) south of Puerto Rico and 90 miles (145 km) north of the Venezuelan capital of Caracas. Going back about 20 years, there have been more than 15 reported incidents of small planes crashing, vanishing, or declaring emergencies while flying in the area. Some have disappeared without a trace. Others have remained lost for years. In January 2013, a twin-engine plane traveling from Los Roques to Caracas crashed, killing the crew and all four passengers—including Italian fashion designer Vittorio

Missoni. The wreckage of the plane was not found until late June. A week earlier, the same search crew found the wreckage of a plane that had crashed in 2008 while flying the same route, killing 14 people.

Although crashes are unfortunately certain to continue, improved aviation technologies may help reduce the chances of getting lost. Advanced equipment arguably could have saved Flight 19 from disappearing. Very high frequency omnidirectional radio range (VOR) was brought into use in the late 1940s and '50s, not long after the Flight 19 incident. The system uses a combination of radio transmissions from the ground and a receiver in a plane to determine the aircraft's course, or its direction in relation to north. VOR is now the world's standard navigational system for airplanes. Its basic structure involves a network of about 3,000 ground stations around the globe, each transmitting directional signals continuously. VOR depends on a plane being within sight of a transmitter, and for small, low-flying planes that might mean a range of only 25 nautical miles (46.3 km). For Flight 19, which reported flying at an altitude of 2,300 feet (701 m), the range would be 40 nautical miles (74.1 km). That's not very far, and VOR stations are only on land, but had VOR been in place in the Bahamas, Flight 19 might have received a reliable directional signal after one of the critical turns on its route.

Nasty Weather

Bermuda is a place associated with warm breezes, sunshine, and shorts. And although Bermuda Triangle mysteries commonly include clear skies, the region is infamous for extremely challenging conditions that can change rapidly. The area is a sort of hurricane highway. Storms that originate off the coast of Africa and head west across the Atlantic most often bend northward, right through the Bermuda Triangle. (From 2000 to 2017, nearly five hurricanes or tropical storms crossed it each year.) The Triangle is also stroked by the Gulf Stream, an ocean current that carries warm water northeast from the Gulf of Mexico all the way to northern Europe. When winds from the northeast blow against the current, the clash of forces can whip up waves high enough to capsize or even split large ships. Similarly, the collision of warm southern air with cold northern air in the winter can set off tremendous storms. The Triangle is also known for "rogue" waves—massive, solitary waves wandering the sea or generated by undersea earthquakes—that can send ships to sudden doom. Then there are waterspouts—seagoing tornadoes that can tear to bits anything unlucky enough to be in their narrow paths.

The VOR network is expensive, involving buildings and other physical equipment that require maintenance and repairs. So despite its long-lasting dependability, VOR is being increasingly replaced by the 24 satellites in the Global Positioning System (GPS), which is much cheaper to operate. GPS emerged in the mid-1990s. It works as satellite signals converge on a receiver on Earth. A car, a boat, a hiker—anyone or anything can carry that receiver. The receiver calculates the difference in time it took for signals to reach it from various satellites, and then uses that information to determine precisely where it is on the surface of the Earth.

Satellite information can also help pilots determine their altitude and speed. That might have been particularly useful for Flight 19. Investigators determined that because Lieutenant Taylor repeatedly asked for

The floor of the Puerto Rico Trench, off the northern coast of Puerto Rico, is 27,480 feet (8,376 m) below the surface.

the time when he was within radio contact, he must not have been wearing a watch. Without knowing how long the planes had been traveling in any direction, he also wouldn't have known how far.

New undersea exploring devices could unlock some of the Bermuda Triangle's secrets as well. But there's a significant obstacle: the area includes some of the deepest reaches in the Atlantic Ocean. The floor of the Puerto Rico Trench, off the northern coast of Puerto Rico, is 27,480 feet (8,376 m) below the surface. Rugged shoals and reefs mark the area, and even in shallower waters, remains that haven't yet been swept off by ocean currents

have often baffled searchers. Many were excited when five Avenger bombers were found in a cluster on the sea bottom 12 miles (19.3 km) off Fort Lauderdale in 1991. Was it Flight 19? Using a remote-controlled submersible camera, scientists found numbers on the planes that led to an unlikely conclusion: the planes all fell into the sea in the same place but at different times and as a result of different accidents. (Evidence suggests crews intentionally ditched the planes.) Investigators instead believe that Flight 19 went down 200 miles (322 km) east of Daytona Beach, which would be about 250 miles (402 km) from where the other Avengers were found.

In 1975, Mike Sibley was a young sailor on a boat near the Bahamas when, early one morning, the sky lit up with a light green glow. "The compass was spinning. I thought we were headed into a parallel universe," said

Sibley. Soon, however, the crew found out from the radio that the air force was conducting a test in the area, using a fluorescent gas that disrupted instruments. Sibley, who later became the chief operating officer of the Grand Yacht Club of Fort Lauderdale, said plane and boat mysteries in the Bermuda Triangle have dwindled in the face of detailed investigations. But he occasionally found himself pulled into one. Compasses on his yachts still spun crazily from time to time. On some trips he got so disoriented by time and the unchanging surroundings at sea that he doubted his own instruments—a common and sometimes fatal flaw for navigators. In 2005, during a charter fishing expedition several miles off the Florida coast from Fort Lauderdale, Sibley rushed to the scene of a small plane crash nearby. Though he arrived soon after the crash and was able to rescue all six people, there were no visible remains of the plane—not so much as an oil slick—at the surface.

The Bermuda Triangle is good for business, Sibley says. It keeps tourists curious about the region. In fact, one of his company's featured outings is the four-day, three-night "Bermuda Triangle," which includes at least one night anchored at sea. Have any clients ever been enveloped in sudden, eggnog-like fogs, or found the sea around them rising on a frothing bubble of methane? "Nothing but beautiful stars," Sibley comments.

The Bermuda Triangle remains one of the busiest shipping and boating areas in the world, crisscrossed by thousands of planes each year as well. But the few that encounter things that can't be explained continue to overshadow the region's routine travel, illustrating the enduring power of mysteries.

Selected Bibliography

Belanger, Jeff. *The Mysteries of the Bermuda Triangle.* New York: Grosset & Dunlap, 2010.

Berlitz, Charles. *The Bermuda Triangle: An Incredible Saga of Unexplained Disappearances.* Garden City, N.Y.: Doubleday, 1974.

Gaddis, Vincent H. "The Deadly Bermuda Triangle." *Argosy* (February 1964): 28–29, 116–18.

Howell, Elizabeth. "Navstar: GPS Satellite Network." SPACE .com. April 27, 2018. https://www.space.com/19794-navstar .html.

Innes, Brian. *The Bermuda Triangle.* Austin, Texas: Steck-Vaughn, 1999.

Kusche, Larry. *The Bermuda Triangle Mystery Solved.* Amherst, N.Y.: Prometheus Books, 1986.

"The 'Mystery' of the Bermuda Triangle." The Museum of UnNatural Mystery. http://www.unmuseum.org/triangle.htm.

Oxlade, Chris. *The Mystery of the Bermuda Triangle.* Des Plaines, Ill.: Heinemann Library, 1999.

Quasar, Gian J. *Into the Bermuda Triangle: Pursuing the Truth Behind the World's Greatest Mystery.* New York: McGraw-Hill, 2004.

Stewart, Gail B. *The Bermuda Triangle.* San Diego: Reference Point, 2009.

Endnotes

air traffic controllers — airport workers who watch radar images of planes approaching or taking off, and communicate with pilots to guide them

altitude — the vertical elevation, or height of an object or point in relation to sea level or ground level

brigantine — a type of two-masted sailing ship

cadets — military or police trainees

extraterrestrial — from beyond Earth

humidity — the amount of invisible water vapor in the air

hype — exaggerated claims or publicity

latitudes — imaginary, horizontal lines on the globe that define location north or south of the equator

magnetic fields — areas in which objects are attracted or repelled by a magnet

meridian — one of many imaginary lines on the earth's surface (actually circles running through the north and south poles) that determine location east and west on the globe

methane — a gas produced by decaying organic matter, such as seaweed

mutiny — a rebellion on a ship, in which the crew seizes command from the captain

mythical	legendary, characteristic of myths (traditional stories that try to explain how something came to be or involve people or things with exaggerated qualities)
nautical miles	distance measurements equal to 1.15 miles (1.85 km), or the circumference of the Earth divided by 21,600
navigation	the use of compasses, maps, charts, and other devices, as well as the sun and stars, to determine one's position or route
parallel universe	a reality that coexists with one's own
phenomena	occurrences that can be observed
radar	a system used to detect objects such as aircraft and determine their position and velocity, derived from the words "radio detection and ranging"
schooner	a type of sailing ship with two or more masts, used for a variety of purposes and favored for its speed
smuggling	illegal transportation of goods or people
submersible	a vessel that can work deep under water, like a submarine
ufologists	people who study reports, recordings, and other evidence related to UFOs
vortices	areas where air (or a liquid or other substance) spins around a center, as in tornadoes or above a tub drain

waterspout	a rotating column or tornado of water and spray formed by a whirlwind
World War I	the war fought in Europe from 1914 to 1918 between the Allied Powers (France, Russia, Great Britain, Italy, and the U.S.) and the Central Powers (Austria-Hungary, Germany, Bulgaria, and the Ottoman Empire)
yachts	recreational boats or ships